Practical Guide to Raising Chickens at Home

James Drive

CONTENTS

Introduction: A Practical Guide to Raising Chickens at Home

Welcome to the captivating world of backyard chicken keeping! In this practical guide, we embark on an exciting journey together to explore the joys and rewards of raising chickens right in the comfort of your own home. Whether you're a seasoned homesteader, a city dweller dreaming of sustainable living, or simply a curious soul looking to connect with nature, this book is your go-to resource for all things chicken.

Why Raise Chickens at Home?

You might be wondering, "Why chickens? Aren't they just farm animals?" Well, dear reader, chickens are much more than just your average barnyard inhabitants. These feathered friends bring a wealth of benefits to your life and the environment. First and foremost, raising chickens provides a direct connection to the source of our food, allowing us to witness the magic of egg production and, if you choose, meat production from start to finish. The satisfaction of gathering fresh, nutritious eggs each morning cannot be overstated!

But that's not all; chickens are incredibly efficient pest controllers, eagerly gobbling up insects and grubs that may threaten your garden. Moreover, they're excellent composters, turning kitchen scraps and yard waste into rich, dark gold that will rejuvenate your garden soil. It's a beautiful cycle of sustainability that mirrors the cycles of life we often overlook in our fast-paced world.

The Ethical and Practical Considerations

Before we delve into the nitty-gritty of chicken keeping, it's important to address some ethical and practical aspects of this endeavor. We firmly believe in responsible and compassionate animal husbandry. Ensuring that your feathered friends have ample space to roam, clean water to drink, nutritious food to eat, and protection from predators is the key to their well-being. We'll explore different coop designs and learn how to create a safe, cozy haven for our chickens.

Additionally, as chicken keepers, we must respect our local regulations and neighbors. Though chickens can thrive in urban, suburban, and rural settings, it's crucial to know the laws regarding backyard chickens in your area. Responsible chicken keeping fosters harmony with the community and showcases the positive aspects of urban agriculture.

Getting Started: The First Steps

Now that we've discussed the "why" and the "how" from a broader perspective, it's time to roll up our sleeves and dive into the nitty-gritty of setting up your home chicken paradise. In the chapters ahead, we'll guide you through selecting the perfect spot for your coop, choosing the right chicken breeds for your needs and preferences, and building or buying a suitable coop and run.

Moreover, we'll discuss the essentials of chicken nutrition, because a well-fed chicken is a happy chicken! You'll learn about the various feed options available and how to supplement their diet with kitchen scraps and garden greens, making your chickens truly part of the family.

A Happy and Healthy Flock

Just like any other pet, chickens require regular care and attention to thrive. In this guide, we'll explore how to recognize signs of good health in your flock, as well as common health issues and how to prevent and address them. Regular health checks and preventive measures will keep your feathered friends clucking contentedly for years to come.

Beyond Eggs: Caring for Chicks and Preparing for Meat Production

While many chicken keepers primarily enjoy the daily supply of farm-fresh eggs, some may venture into the exciting world of raising chicks. We'll discuss the magical experience of incubation, caring for the adorable fluffballs, and integrating them into your existing flock. If you're interested in exploring meat production, we'll also guide you through the ethical considerations and the processes involved.

Join Us on this Feathered Adventure!

Whether you're seeking a deeper connection with your food, a greener approach to life, or simply a joyful addition to your family, raising chickens at home is a fulfilling and enchanting pursuit. Throughout this guide, we'll share practical advice, heartwarming stories, and tried-and-true tips from experienced chicken keepers.

So, fellow poultry enthusiasts, let's embark on this feathered adventure together! By the end of this book, you'll be well-equipped to create a haven for your flock, enjoy the delightful company of these charming creatures, and revel in the wonder of sustainable living. The world of home chicken keeping awaits, and we couldn't be more excited to share it with you! Happy clucking!

SETTING UP THE AREA

<u>I</u>ntroduction

Setting up the area for your backyard chicken farming venture is a crucial step in ensuring the health and well-being of your feathered friends. Creating a safe and comfortable environment is essential to their overall happiness and productivity. In this chapter, we will explore the key considerations for choosing the right location, constructing a suitable coop and run, and providing the necessary amenities for your chickens. <h2>Choosing the Location and Size</h2> Selecting the right location for your chicken coop and run is vital for the success of your home chicken raising project. Consider these factors when determining the perfect spot:

1. **Sunlight Exposure:** Optimal sunlight exposure is essential for the chickens' health and egg production. Choose an area with ample sunlight during the day.

2. **Drainage:** Ensure the chosen location has good drainage to prevent water-logging during rainy periods. Standing water can lead to health issues for the chickens.

3. **Proximity to Your Home:** Placing the coop within easy reach of your home will make daily tasks more convenient, such as feeding, watering, and egg collection.

4. **Predator Protection:** Position the coop away from dense vegetation or tall structures that could provide hiding spots for predators. Additionally, consider using predator-proof materials and measures to safeguard your chickens.

<u>Constructing a Suitable Coop and Run</u>

Now that you've chosen the location, it's time to build a chicken coop and run that suits the needs of your flock. Here are some important factors to consider:

1. **Coop Design:** The coop should provide adequate space for your chickens to roost comfortably at night and lay eggs in nest boxes during the day. A well-designed coop allows for easy cleaning and maintenance.

2. **Run Area:** The run is an outdoor space where your chickens can freely roam, scratch, and peck. Provide at least 8-10 square feet of run space per chicken to prevent overcrowding.

3. **Materials and Ventilation:** Use sturdy and predator-resistant materials for the coop, such as hardware cloth and pressure-treated wood. Proper ventilation is essential to prevent moisture buildup and promote good air quality.

4. **Nesting Boxes:** Incorporate nesting boxes inside the coop to encourage egg-laying in a clean and safe environment. Aim for one nesting box per 3-4 hens.

<u>Providing Necessary Amenities</u>

Happy and healthy chickens require more than just food and shelter. Here are some essential amenities to include in your chicken setup:

1. **Feeding and Watering Stations:** Place feeders and waterers inside the coop and run to ensure easy access to food and clean water throughout the day.

2. **Dust Bath Area:** Chickens love to take dust baths to keep their feathers clean and free from parasites. Create a designated dust bath area using sand and diatomaceous earth.

3. **Perches and Roosts:** Install sturdy perches and roosts inside the coop at varying heights to allow chickens to rest comfortably at night.

4. **Enrichment Items:** Provide toys and objects for chickens to peck at, such as hanging cabbage or treat dispensers. This helps keep them entertained and reduces stress.

<u>Examples of Coop Layout and Design</u>

To help you visualize the process of setting up the area for your chickens, let's explore two different examples of coop layouts:

Example 1: Urban Backyard Coop

- Location: A sunny corner of the backyard with easy access to the kitchen for daily chores.

- Coop Design: A small and compact A-frame coop with two nesting boxes and a removable tray for easy cleaning. It includes a small attached run with a wire mesh top for predator protection.

- Amenities: Perches and a dust bath area are included inside the coop. The waterer and feeder are placed both inside the coop and in the run for convenience.

Example 2: Suburban Homestead Coop

- Location: A spacious area at the edge of the property with plenty of sunlight and a view of the garden.

- Coop Design: A larger, stationary coop with a walk-in design, providing ample space for up to 10 chickens. The coop has multiple nesting boxes, a window for natural light and ventilation, and predator-resistant features.

- Run Area: A sizable run attached to the coop with wire mesh sides and a roof, allowing chickens to forage freely while staying protected.

- Amenities: Alongside perches and a dust bath area, the setup includes hanging cabbage, treat dispensers, and a grazing area with fresh greens for added enrichment.

<u>Planning for Expansion</u>

As you embark on your chicken-raising journey, it's essential to plan for potential expansion. Whether you initially start with a small flock or a larger one, consider the following tips for future growth:

1. **Scalable Coop Design:** Build a coop that can accommodate a larger number of chickens if you plan to expand your flock. Design the coop with modular

features to add nesting boxes or additional roosts easily.

2. **Expandable Run:** Create a run area that can be extended or reconfigured as your flock grows. Consider using portable electric netting to create temporary grazing areas for rotational grazing.

3. **Zoning and Regulations:** Check local zoning laws and regulations regarding the maximum number of chickens allowed on your property. Ensure your expansion plans comply with the rules.

In conclusion, setting up the area for your chickens involves thoughtful planning, proper coop and run design, and attention to the chickens' well-being. Examples of coop layouts and considerations for future expansion can guide you in creating a successful and rewarding backyard chicken farming experience.

<u>Example of Coop Layout and Design</u>

To help you visualize the process of setting up the area for your chickens, let's explore two different examples of coop layouts:

Example 1: Urban Backyard Coop

- Location: A sunny corner of the backyard with easy access to the kitchen for daily chores.

- Coop Design: A small and compact A-frame coop with two nesting boxes and a removable tray for easy cleaning. It includes a small attached run with a wire mesh top for predator protection.

- Amenities: Perches and a dust bath area are included inside the coop. The waterer and feeder are placed both inside the coop and in the run for convenience.

Example 2: Suburban Homestead Coop

- Location: A spacious area at the edge of the property with plenty of sunlight and a view of the garden.

- Coop Design: A larger, stationary coop with a walk-in design, providing ample space for up to 10 chickens. The coop has multiple nesting boxes, a window for natural light and ventilation, and predator-resistant features.

- Run Area: A sizable run attached to the coop with wire mesh sides and a roof, allowing chickens to forage freely while staying protected.

- Amenities: Alongside perches and a dust bath area, the setup includes hanging cabbage, treat dispensers, and a grazing area with fresh greens for added enrichment.

Planning for Expansion

As you embark on your chicken-raising journey, it's essential to plan for potential expansion. Whether you initially start with a small flock or a larger one, consider the following tips for future growth:

1. **Scalable Coop Design:** Build a coop that can accommodate a larger number of chickens if you plan to expand your flock. Design the coop with modular features to add nesting boxes or additional roosts easily.

2. **Expandable Run:** Create a run area that can be extended or reconfigured as your flock grows. Consider using portable electric netting to create temporary grazing areas for rotational grazing.

3. **Zoning and Regulations:** Check local zoning laws and regulations regarding the maximum number of chickens allowed on your property. Ensure your expansion plans comply with the rules.

In conclusion, setting up the area for your chickens involves thoughtful planning, proper coop and run design, and attention to the chickens' well-being. Examples of coop layouts and considerations for future expansion can guide you in creating a successful and rewarding backyard chicken farming experience.

Choosing the Location and Size

Selecting the right location and determining the appropriate size for your chicken coop and run are crucial steps in creating a suitable environment for your chickens. This subchapter will explore the key factors to consider when making these decisions, with practical examples to guide you.

- **Sunlight Exposure:** Proper sunlight exposure is essential for the overall health and productivity of your chickens. When choosing the location, look for an area that receives ample sunlight throughout the day. Consider examples of suitable locations, such as south-facing backyards or open spaces with minimal shading.

- **Drainage Considerations:** Ensuring good drainage in the chosen area is vital to prevent waterlogging during rainy periods. Discuss the importance of proper drainage to avoid health issues for the chickens. Provide examples of ways to improve drainage, such as creating a slight slope or adding gravel to the ground.

- **Proximity to Your Home:** Placing the coop and run within easy reach of your home offers several advantages. Discuss the benefits of having the coop close to the house, such as convenient access for feeding, cleaning, and monitoring. Offer examples of how this proximity can streamline daily chicken care tasks.

- **Predator Protection:** Predators can pose a significant threat to your chickens' safety. Highlight the importance of predator-proofing the coop and run area. Provide examples of predator-resistant materials, such as hardware cloth, and measures like burying wire mesh to deter digging predators.

<u>Constructing a Suitable Coop and Run</u>

Building a well-designed coop and run is essential for your chickens' safety and well-being. In this subchapter, we'll delve into the key considerations when constructing your coop and run, along with practical examples to guide you.

- **Coop Design Essentials:** The coop's design plays a crucial role in the overall comfort and health of your chickens. Discuss the elements of a well-designed coop, such as proper ventilation, insulation, and ease of cleaning. Provide examples of coop designs that meet these criteria, including detailed diagrams.

- **Run Area Size and Features:** The run area is where your chickens will spend a significant amount of their time. Explain the importance of providing enough space for them to forage and exercise. Offer examples of how to calculate the appropriate run size based on the number of chickens and recommended guidelines.

- **Materials and Safety:** The materials used in coop and run construction should be sturdy and safe for your chickens. Discuss the importance of using predator-resistant materials and how they enhance the safety of your flock. Offer examples of different materials and their benefits.

- **Nesting Boxes and Egg Collection:** Proper nesting boxes are essential for encouraging egg-laying in a clean and comfortable environment. Explain the necessity of nesting boxes and their proper placement inside the coop. Offer examples of different nesting box designs and how to ensure easy access for egg collection.

<u>Providing Necessary Amenities</u>

Meeting the basic needs of your chickens is vital for their overall health and happiness. In this subchapter, we'll explore the essential amenities your chicken area should have, along with practical examples to assist you.

- **Feeding and Watering Stations:** Adequate feeding and watering stations are necessary to ensure your chickens have access to food and clean water at all times. Discuss the importance of providing accessible feeding and watering stations. Offer examples of different feeder and waterer designs suitable for your setup.

- **Dust Bath Area:** Chickens love to take dust baths as part of their grooming routine. Explain the benefits of a dust bath area for your chickens' health and how it helps control external parasites. Provide examples of how to create a designated dust bath spot using sand and diatomaceous earth.

- **Perches and Roosts:** Providing perches and roosts inside the coop is essential for your chickens' comfort and safety at night. Discuss the significance of different perch designs and how to accommodate different-sized chickens. Offer examples of perch and roost placements to maximize space and prevent crowding.

- **Enrichment Items:** Keeping your chickens mentally and physically stimulated is crucial for their well-being. Discuss the importance of providing toys and enrichment items to reduce boredom and stress. Offer examples of hanging cabbage, treat dispensers, and other enriching activities to keep your chickens happy and entertained.

SELECTING THE BREEDS

When embarking on the journey of raising chickens at home, choosing the right breeds is a crucial decision that will impact your overall experience and the flock's success. Each chicken breed comes with its unique characteristics, temperament, and purpose. In this chapter, we will explore the different factors to consider when selecting breeds, and provide practical examples and a comparison table to assist you in making an informed decision.

<u>Factors to Consider</u>

When choosing chicken breeds for your home flock, several factors should guide your decision-making process. Let's examine the key considerations:

Egg Production: If your primary goal is egg production, consider breeds known for their prolific laying capabilities. Examples include the Leghorn, Rhode Island Red, and Australorp. They can produce a high number of eggs per year, making them ideal for egg-centric flocks.

Meat Production: For those interested in raising chickens for meat, broiler breeds like the Cornish Cross and the Freedom Ranger are popular choices. These breeds are specifically bred for fast growth and superior meat quality.

Dual-Purpose Breeds: Some breeds, like the Plymouth Rock and Sussex, are known as dual-purpose breeds. They are well-balanced in terms of egg production and meat quality, making them suitable for both purposes.

Temperament: Consider the temperament of the breeds you are interested in. Some breeds are docile and friendly, making them ideal for families and small children. Others might be more flighty or independent, which may impact their suitability for a home setting.

Climate Adaptability: Certain breeds are better suited for specific climates. For instance, Mediterranean breeds like the Leghorn tolerate heat well, while cold-hardy breeds like the Orpington fare better in colder regions.

Space Requirements: Some breeds thrive in confined spaces, while others require more room to roam. Factor in the available space in your backyard when selecting breeds.

Breeds Comparison Table

To help you compare the characteristics of different chicken breeds, here's a table outlining some popular breeds and their primary attributes:

Breed	Egg Production	Meat Production	Temperament	Climate Adaptability	Space Requirements
Leghorn	Excellent	Fair	Flighty	Heat-tolerant	Confined space
Rhode Island Red	Good	Fair	Docile	All climates	Moderate space
Australorp	Excellent	Fair	Friendly	Heat-tolerant	Moderate space
Cornish Cross	Poor	Excellent	Calm	All climates	Confined space
Freedom Ranger	Fair	Excellent	Docile	All climates	Moderate space
Plymouth Rock	Good	Good	Friendly	All climates	Moderate space
Sussex	Good	Good	Calm	All climates	Moderate space
Orpington	Good	Fair	Friendly	Cold-hardy	Moderate space

To further illustrate the breed selection process, let's consider two different scenarios:

Scenario 1: Egg-Centric Backyard For a family looking to enjoy a steady supply of eggs, the Leghorn and Australorp breeds could be excellent choices. The Leghorn, with its exceptional egg production, suits a smaller space, while the friendly Australorp complements the flock with its calm temperament.

Scenario 2: Homestead with Dual Purpose For a homestead seeking both eggs and meat, the Plymouth Rock and Freedom Ranger breeds offer a balanced approach. The Plymouth Rock provides a reliable egg supply, and the Freedom Ranger ensures a quality meat source when needed.

Egg Production Breeds

When focusing on egg production, certain chicken breeds excel in consistently laying a high number of eggs. Let's explore some popular egg-centric breeds and their key characteristics:

Leghorn:

- Egg Production: Leghorns are renowned for their outstanding egg-laying abilities, producing around 280-320 large white eggs per year.

- Temperament: Leghorns are active and flighty, making them excellent foragers. They are more independent and might not be as affectionate as other breeds.

- Climate Adaptability: These Mediterranean breeds handle heat well and thrive in warm climates.

- Space Requirements: Leghorns do well in confined spaces and are a good choice for smaller backyards or urban settings.

Rhode Island Red:

- Egg Production: Rhode Island Reds are dependable layers, providing approximately 200-300 large brown eggs per year.

- Temperament: They have a calm and friendly demeanor, making them suitable for families with children.

- Climate Adaptability: Rhode Island Reds adapt well to various climates, making them a versatile choice.

- Space Requirements: These birds do well with moderate space and can thrive in both confined and free-range setups.

Australorp:

- Egg Production: Australorps are prolific layers, producing about 250-300 large brown eggs per year.

- Temperament: Known for their docile and friendly nature, Australorps are great additions to a family flock.

- Climate Adaptability: They tolerate heat well, making them suitable for warmer regions.

- Space Requirements: Australorps do well with moderate space and can adapt to various coop sizes.

Meat Production Breeds

For those interested in raising chickens primarily for meat, certain breeds are specifically bred for their superior meat quality and fast growth. Let's explore some popular meat-centric breeds and their key characteristics:

Cornish Cross:

- Meat Production: Cornish Cross chickens grow rapidly and reach market weight within 6-8 weeks. They are the standard breed for commercial meat production.

- Temperament: These birds are generally calm and docile.

- Climate Adaptability: Cornish Cross chickens adapt well to different climates.

- Space Requirements: Due to their rapid growth, they are best suited for a confined space environment.

Freedom Ranger:

- Meat Production: Freedom Rangers are known for their flavorful meat and slower growth rate compared to Cornish Cross. They reach market weight within 9-11 weeks.

- Temperament: These birds have a docile and friendly nature.

- Climate Adaptability: Freedom Rangers adapt well to various climates.

- Space Requirements: They do well with moderate space and can thrive in both confined and free-range environments.

Dual-Purpose Breeds

Dual-purpose breeds are versatile additions to a home flock, offering a balance of egg production and meat quality. Let's explore some popular dual-purpose breeds and their key characteristics:

Plymouth Rock:

- Egg Production: Plymouth Rocks are reliable layers, providing approximately 200-280 large brown eggs per year.

- Meat Production: They have good meat quality, making them suitable for both egg and meat production purposes.

- Temperament: These birds are friendly and easy to handle.

- Climate Adaptability: Plymouth Rocks adapt well to various climates.

- Space Requirements: They do well with moderate space and can adapt to different coop sizes.

Sussex:

- Egg Production: Sussex chickens are good layers, providing around 200-250 medium to large brown eggs per year.

- Meat Production: They have decent meat quality, making them suitable for both egg and meat production purposes.

- Temperament: Sussex chickens have a calm and friendly demeanor.

- Climate Adaptability: They handle various climates well, including both heat and cold.

- Space Requirements: Sussex chickens do well with moderate space and can adapt to different coop sizes.

Orpington:

- Egg Production: Orpingtons are good layers, producing approximately 180-200 large brown eggs per year.

- Meat Production: They have good meat quality, making them suitable for both egg and meat production purposes.

- Temperament: Orpingtons are friendly and docile, making them great additions to a family flock.

- Climate Adaptability: They are cold-hardy and can handle colder regions well.

- Space Requirements: Orpingtons do well with moderate space and can adapt to different coop sizes.

In conclusion, selecting the right chicken breeds for your home flock involves considering various factors such as egg or meat production, temperament, climate adaptability, and available space. By comparing different breeds and understanding their unique attributes, you can make an informed decision that aligns with your specific goals and preferences.

FEEDING AND NUTRITION

Feeding and nutrition play a vital role in the health, growth, and overall well-being of your backyard chickens. Providing a balanced diet with the right nutrients is essential for supporting their egg production, meat development (if applicable), and overall vitality. In this chapter, we will delve into the basics of a balanced chicken diet, discuss the options between commercial feeds and homemade feeds, and explore how to manage feed according to the chickens' age and life stages.

Basics of a Balanced Chicken Diet

A well-balanced chicken diet consists of various nutrients necessary for their growth and maintenance. The basic components of a chicken's diet include:

- **Protein:** Protein is crucial for muscle development, egg production, and feather growth. Chickens need a diet with sufficient protein levels, especially during their growing stages and laying periods.

- **Carbohydrates:** Carbohydrates provide energy for chickens to carry out their daily activities. Grains and other carbohydrate sources help meet their energy needs.

- **Fats:** Fats are an essential energy source and play a role in maintaining healthy skin and feathers. Including fats in their diet is especially important during colder months.

- **Vitamins and Minerals:** Chickens require various vitamins and minerals for overall health and proper bodily functions. These include vitamins A, D, E, K, as well as calcium, phosphorus, and others.

- **Water:** Access to clean and fresh water is critical at all times. Chickens require water for digestion, thermoregulation, and general hydration.

Commercial Feeds vs. Homemade Feeds

There are two primary options for providing your chickens with their diet: commercial feeds and homemade feeds. Both have their advantages and disadvantages.

Commercial Feeds:

- Convenience: Commercial feeds are readily available and save time in preparing a balanced diet.

- Nutrient Balance: These feeds are formulated to meet the nutritional needs of chickens at different life stages.

- Variety: There are various commercial feeds, such as starter, grower, layer, and broiler feeds, tailored for specific purposes.

- Cost: While convenient, the cost of commercial feeds may be higher, especially for larger flocks.

Homemade Feeds:

- Customization: Preparing homemade feeds allows you to control the ingredients and customize the diet for your chickens.

- Natural Ingredients: Some chicken owners prefer using organic or non-GMO ingredients in homemade feeds.

- Cost: Homemade feeds can be cost-effective, especially if you can source ingredients in bulk.

- Nutrient Balance: Preparing a balanced homemade diet requires careful planning and knowledge of the chickens' nutritional needs.

A combination of both commercial feeds and homemade supplements can be a good approach. Commercial feeds can serve as the base diet, supplemented with kitchen scraps, garden produce, and occasional treats to provide variety and additional nutrients.

<u>Managing Feed According to Age and Life Stages</u>

Chickens have different nutritional requirements at various stages of their lives. Understanding how to manage their feed according to their age and life stages is essential for their health and productivity.

Chick Starter (0-8 weeks):

- High Protein: Chicks require a higher protein content (18-20%) to support their rapid growth and development.

Grower (8-16 weeks):

- Lower Protein: As chicks mature, the protein content can be reduced to around 16-18% to support steady growth without excessive weight gain.

Layer (16+ weeks):

- Calcium: Layers need a diet with added calcium (about 3.5-4%) to support eggshell formation.

- Lower Protein: The protein content can be reduced to around 15-16% for adult layers.

Broiler (Meat) Chickens:

- High Protein: Broilers require a high-protein diet (20-23%) to encourage fast and efficient muscle development.

Molting Hens:

- High Protein and Fats: During molting, hens benefit from increased protein and fats to support feather regrowth.

It is essential to monitor your chickens' body condition and adjust their diet as needed. Providing clean, fresh water at all times is critical, especially during hot weather or when feeding dry diets.

Feeding and nutrition play a vital role in the health, growth, and overall well-being of your backyard chickens. Providing a balanced diet with the right nutrients is essential for supporting their egg production, meat development (if applicable), and overall vitality. In this chapter, we will delve into the basics of a balanced chicken diet, discuss the options between commercial feeds and homemade feeds, and explore how to manage feed according to the chickens' age and life stages.

<u>Basics of a Balanced Chicken Diet</u>

A well-balanced chicken diet consists of various nutrients necessary for their growth and maintenance. The basic components of a chicken's diet include:

Protein: Protein is crucial for muscle development, egg production, and feather growth. Chickens need a diet with sufficient protein levels, especially during their growing stages and laying periods.

Carbohydrates: Carbohydrates provide energy for chickens to carry out their daily activities. Grains and other carbohydrate sources help meet their energy needs.

Fats: Fats are an essential energy source and play a role in maintaining healthy skin and feathers. Including fats in their diet is especially important during colder months.

Vitamins and Minerals: Chickens require various vitamins and minerals for overall health and proper bodily functions. These include vitamins A, D, E, K, as well as calcium, phosphorus, and others.

Water: Access to clean and fresh water is critical at all times. Chickens require water for digestion, thermoregulation, and general hydration.

<u>Commercial Feeds vs. Homemade Feeds</u>

There are two primary options for providing your chickens with their diet: commercial feeds and homemade feeds. Both have their advantages and disadvantages.

Commercial Feeds:

Convenience: Commercial feeds are readily available and save time in preparing a balanced diet.

Nutrient Balance: These feeds are formulated to meet the nutritional needs of chickens at different life stages.

Variety: There are various commercial feeds, such as starter, grower, layer, and broiler feeds, tailored for specific purposes.

Cost: While convenient, the cost of commercial feeds may be higher, especially for larger flocks.

Homemade Feeds:

Customization: Preparing homemade feeds allows you to control the ingredients and customize the diet for your chickens.

Natural Ingredients: Some chicken owners prefer using organic or non-GMO ingredients in homemade feeds.

Cost: Homemade feeds can be cost-effective, especially if you can source ingredients in bulk.

Nutrient Balance: Preparing a balanced homemade diet requires careful planning and knowledge of the chickens' nutritional needs.

A combination of both commercial feeds and homemade supplements can be a good approach. Commercial feeds can serve as the base diet, supplemented with kitchen scraps, garden produce, and occasional treats to provide variety and additional nutrients.

Managing Feed According to Age and Life Stages

Chickens have different nutritional requirements at various stages of their lives. Understanding how to manage their feed according to their age and life stages is essential for their health and productivity.

Chick Starter (0-8 weeks):

High Protein: Chicks require a higher protein content (18-20%) to support their rapid growth and development.

Grower (8-16 weeks):

Lower Protein: As chicks mature, the protein content can be reduced to around 16-18% to support steady growth without excessive weight gain.

Layer (16+ weeks):

Calcium: Layers need a diet with added calcium (about 3.5-4%) to support eggshell formation.

Lower Protein: The protein content can be reduced to around 15-16% for adult layers.

Broiler (Meat) Chickens:

High Protein: Broilers require a high-protein diet (20-23%) to encourage fast and efficient muscle development.

Molting Hens:

High Protein and Fats: During molting, hens benefit from increased protein and fats to support feather regrowth.

It is essential to monitor your chickens' body condition and adjust their diet as needed. Providing clean, fresh water at all times is critical, especially during hot weather or when feeding dry diets.

<u>Supplementing and Treats</u>

Supplementing your chickens' diet with treats and additional supplements can be a rewarding way to keep them happy and healthy. However, it's essential to provide these extras in moderation to avoid nutritional imbalances or obesity.

Kitchen Scraps and Garden Produce:

- Kitchen scraps like vegetable peels, fruit leftovers, and bread can be a great addition to your chickens' diet. However, avoid feeding them spoiled or moldy food.

- Fresh garden produce, such as lettuce, kale, and zucchini, can be given as treats. Chickens enjoy foraging for greens and bugs in the garden, which also provides additional enrichment.

Mealworms and Insects:

- Mealworms and other insects are an excellent source of protein for chickens. They love hunting for insects in the coop or run, which promotes natural behaviors.

Calcium Supplements:

- Providing additional calcium sources, such as crushed eggshells or oyster shells, is crucial for laying hens to maintain strong eggshells. Place these supplements in a separate container for free-choice consumption.

Probiotics and Vitamins:

- Probiotics can be added to the water or feed to promote a healthy gut flora in chickens.

- Vitamin supplements may be necessary in certain situations, such as when chickens are under stress or experiencing health issues. Consult with a veterinarian for specific guidance.

<u>Feeding Management</u>

Managing your chickens' feeding routine involves more than just providing food. Proper feeding management ensures that your chickens receive the right amount of feed, prevents wastage, and encourages natural behaviors.

Free-Choice Feeding vs. Time-Restricted Feeding:

- Free-choice feeding allows chickens to eat as much as they want throughout the day. It's suitable for non-laying chickens or when using commercial feeds designed for free-choice consumption.

- Time-restricted feeding is more common for laying hens to control their diet and prevent excessive weight gain. Offer feed twice a day, and remove any uneaten feed after 20-30 minutes.

Feeder and Waterer Placement:

- Position feeders and waterers at a comfortable height for chickens to access easily.

- Place them in a sheltered area to protect the feed from rain and prevent contamination.

Regular Cleaning and Maintenance:

- Clean feeders and waterers regularly to prevent the growth of harmful bacteria and mold.

- Remove any wet or soiled feed promptly to maintain feed quality.

Monitoring Body Condition:

- Regularly assess your chickens' body condition to ensure they are neither underweight nor overweight.

- Adjust the feed amount accordingly to maintain a healthy body condition.

<u>Seasonal Considerations</u>

Feeding your chickens may require some adjustments based on seasonal changes and weather conditions.

Summer:

- Provide plenty of fresh, cool water to prevent dehydration in hot weather.

- Consider feeding during the cooler parts of the day to reduce heat stress.

Winter:
- Increase the fat content in their diet to help chickens cope with colder temperatures.

- Offer warm oatmeal or cooked grains during cold spells to provide extra warmth.

Molting Season:
- During molting, increase protein levels to support feather regrowth.

- Supplement their diet with vitamins and minerals to promote overall health.

In conclusion, proper feeding and nutrition are fundamental to your chickens' health and productivity. By providing a balanced diet, supplementing with treats in moderation, and practicing effective feeding management, you can ensure the well-being and happiness of your backyard flock year-round.

HEALTH AND DISEASE MANAGEMENT

Maintaining the health of your backyard chickens is crucial for their overall well-being and productivity. Preventive measures and prompt action in case of illness can significantly impact the success of your chicken-raising venture. In this chapter, we will explore the essentials of good health practices, common chicken diseases, and how to manage and prevent health issues in your flock.

<u>Preventive Health Practices</u>

Implementing preventive health practices can help minimize the risk of diseases and maintain a healthy flock. Here are some essential practices to follow:

Biosecurity Measures:

- Limit access to your chicken area to minimize the introduction of diseases.

- Quarantine new birds before integrating them into the existing flock.

Cleanliness and Hygiene:

- Keep the coop and run area clean by regularly removing manure and soiled bedding.

- Disinfect the coop periodically using poultry-safe disinfectants.

Nutrition:

- Provide a well-balanced diet with adequate nutrients to support their immune system.

- Avoid feeding moldy or spoiled food that could lead to health issues.

Ventilation:

- Ensure proper ventilation in the coop to reduce moisture and ammonia buildup.

- Good air circulation helps prevent respiratory issues.

Parasite Control:

- Monitor for external parasites like mites and lice and treat them promptly.

- Administer regular deworming treatments to control internal parasites.

Common Chicken Diseases

Understanding common chicken diseases can help you identify symptoms early and take appropriate action. Here are some diseases to be aware of:

Avian Influenza:

- Highly contagious viral disease affecting the respiratory and digestive systems.

- Symptoms include coughing, sneezing, swollen face, decreased egg production, and sudden death.

- Strict biosecurity measures are crucial to prevent its spread.

Coccidiosis:

- Caused by intestinal protozoa.

- Symptoms include diarrhea, lethargy, and decreased appetite.

- Maintain clean and dry bedding and consider medicated feed to prevent coccidiosis.

Marek's Disease:

- Viral disease affecting the nervous system.

- Symptoms include paralysis, uncoordinated movements, and weight loss.

- Vaccination is recommended for preventing Marek's disease.

Infectious Bronchitis:

- Highly contagious viral disease affecting the respiratory system.

- Symptoms include coughing, sneezing, nasal discharge, and reduced egg production.

- Vaccination can help prevent infectious bronchitis.

Disease Management and Prevention

Prompt action is essential if you suspect a disease in your flock. Here are steps to manage and prevent diseases:

Isolation:

- Isolate sick birds immediately to prevent the spread of disease to the rest of the flock.

- Consult with a veterinarian to determine the appropriate course of action.

Treatment:

- Administer treatments as prescribed by a veterinarian for specific diseases.

- Follow dosage and treatment instructions carefully.

Vaccination:

- Vaccinate your chickens against common diseases based on regional prevalence.

- Regularly update vaccinations according to the recommended schedule.

Observation:

- Regularly observe your chickens for any signs of illness.

- Early detection can lead to timely intervention and better outcomes.

Record Keeping:

- Maintain detailed records of vaccinations, treatments, and any health issues in your flock.

- This information can help track patterns and make informed decisions.

Disease Prevention Vaccination Schedule

Preventive vaccination is a crucial part of disease management in your flock. Here is a vaccination schedule to consider for common chicken diseases:

Disease	Vaccine	Age of Vaccination	Booster Shots
Marek's Disease	HVT, SB-1, or CVI-988	Day-old chicks	None
Infectious Bronchitis	Live or inactivated vaccine	2-3 weeks old	6-8 weeks old
Newcastle Disease	Live or inactivated vaccine	2-3 weeks old	6-8 weeks old
Fowl Pox	Live or inactivated vaccine	8-12 weeks old	None

Note: The vaccination schedule may vary depending on regional disease prevalence and your veterinarian's recommendations.

Addressing Parasite Infestations

Parasite infestations can significantly impact your chickens' health. Regular monitoring and appropriate treatments are essential. Here are common parasites and their management:

External Parasites:

Parasite	Symptoms	Treatment and Prevention
Mites	Feather loss, skin irritation	Apply poultry-safe insecticides to the coop and chickens.
Lice	Feather picking, irritation	Dust chickens with poultry-safe insecticides.

Internal Parasites:

Parasite	Symptoms	Treatment and Prevention
Roundworms	Weight loss, diarrhea	Administer regular deworming treatments.
Tapeworms	Weight loss, pale combs	Administer deworming treatments specific to tapeworms.
Coccidia	Bloody diarrhea	Provide medicated feed or treatments for coccidiosis.

Emergency First Aid Kit

Having an emergency first aid kit on hand can be invaluable in case of injuries or health emergencies. Here's a list of essential items:

Sterile Gauze Pads

Antiseptic Solution

Scissors and Tweezers

Disposable Gloves

Vet Wrap or Bandages

Poultry Electrolytes

Thermometer

Veterinary Contact Information

First Aid Book for Chickens

<u>Quarantine Procedures</u>

When introducing new chickens to your flock, proper quarantine procedures are crucial to prevent disease transmission. Here's how to quarantine new birds:

Isolate new birds in a separate, clean coop and run area.

Observe them for signs of illness for at least 2-3 weeks.

Monitor for symptoms like coughing, sneezing, lethargy, or diarrhea.

Only introduce the new birds to the existing flock after a clean bill of health.

In conclusion, maintaining the health of your backyard flock requires a combination of preventive measures, vigilant observation, and prompt action in case of illness or parasite infestations. Following a vaccination schedule, addressing parasite issues, having an emergency first aid kit, and implementing proper quarantine procedures can help ensure the well-being and longevity of your chickens.

COOP MAINTENANCE AND CLEANING

A clean and well-maintained coop is essential for the health and comfort of your chickens. Regular cleaning and proper coop management contribute to a hygienic environment that supports optimal egg production, reduces the risk of diseases, and enhances the overall well-being of your flock. In this chapter, we will explore the necessary steps for coop maintenance, cleaning routines, and coop design considerations to ensure a happy and healthy home for your chickens.

<u>Coop Design Considerations</u>

When designing or choosing a chicken coop, certain features can make maintenance and cleaning more manageable. Consider the following aspects:

- **Easy Access:** Ensure the coop has sufficient access points, such as large doors or removable panels, to facilitate cleaning and egg collection.

- **Predator-Proofing:** A secure coop design with sturdy materials and properly installed predator-proofing features reduces the risk of pest infestations and predator attacks.

- **Ventilation:** Proper ventilation is essential to reduce moisture buildup and prevent respiratory issues. Incorporate windows or vents with mesh coverings to allow for adequate airflow.

- **Roosts and Nesting Boxes:** Elevate roosts off the ground to keep them clean and provide easy cleaning access underneath. Place nesting boxes in a way that allows for straightforward egg collection and cleaning.

- **Flooring Materials:** Consider using materials like removable trays or linoleum flooring that can be easily cleaned and replaced if necessary.

Regular Cleaning Routine

Establishing a regular cleaning routine for your coop is essential for maintaining a healthy environment. Here's a suggested cleaning schedule:

Daily Tasks:

- Remove any wet bedding or soiled areas to prevent ammonia buildup.

- Collect eggs regularly to ensure cleanliness and freshness.

Weekly Tasks:

- Sweep or rake the coop floor to remove debris and dust.

- Add fresh bedding material as needed.

Monthly Tasks:

- Completely clean and disinfect the coop.

- Remove all bedding material and replace it with fresh, clean bedding.

- Inspect and clean nesting boxes and roosts.

Seasonal Tasks:

- Perform a deep cleaning of the coop before winter and summer to prepare for temperature changes.

Deep Cleaning and Disinfection

Performing a deep cleaning and disinfection of the coop is crucial to eliminate harmful bacteria and parasites. Follow these steps:

1. **Empty the Coop:** Remove all chickens and accessories from the coop.

2. **Remove Bedding and Debris:** Take out all bedding material, feathers, and debris.

3. **Scrub and Wash:** Scrub the coop's interior walls, roosts, and nesting boxes with

a mixture of water and poultry-safe disinfectant.

4. **Allow to Dry:** Allow the coop to air dry completely before adding fresh bedding.

5. **Replace Bedding:** Add clean and dry bedding material to the coop.

6. **Sanitize Accessories:** Clean and disinfect feeders, waterers, and other accessories before returning them to the coop.

7. **Return Chickens:** Once the coop is dry and fresh bedding is added, allow the chickens back inside.

Pest Control

Regular coop cleaning and maintenance are essential for controlling pests. Here are some additional measures to keep pests at bay:

- **Inspect Regularly:** Routinely inspect the coop for signs of pests such as mites, lice, rodents, and other insects.

- **Treat Infestations:** If pests are detected, take immediate action to treat the infestation using poultry-safe insecticides or other appropriate methods.

- **Preventive Measures:** Implement preventive measures, such as keeping the coop area clean and free from food scraps, to discourage pests from entering.

Coop Run Maintenance

If your chickens have access to an outdoor run, maintaining this area is also important. Here are some tips for coop run maintenance:

- **Remove Droppings:** Regularly rake or shovel droppings from the run area to keep it clean and prevent odor buildup.

- **Rotate Access:** Consider rotating the chickens' access to different areas of the run to allow the ground to recover and reduce parasite buildup.

- **Provide Dust Bath:** Offer a designated dust bath area filled with sand and diatomaceous earth to help control external parasites.

In conclusion, proper coop maintenance and cleaning are essential for providing your chickens with a healthy and comfortable living environment. Regular cleaning routines, deep cleaning and disinfection, pest control measures, and maintaining the coop run contribute to a thriving flock and a more enjoyable chicken-raising experience.

CHICKEN BEHAVIOR AND ENRICHMENT

Understanding chicken behavior and providing enrichment activities are essential for keeping your flock mentally stimulated, happy, and well-adjusted. Chickens are intelligent and social animals, and meeting their behavioral needs is crucial for their overall welfare. In this chapter, we will explore common chicken behaviors, the importance of enrichment, and various activities to keep your chickens engaged and content.

<u>Understanding Chicken Behavior</u>

Observing and understanding chicken behavior can provide valuable insights into their well-being and needs. Here are some common chicken behaviors and what they might indicate:

- **Foraging:** Chickens are natural foragers, scratching and pecking at the ground to find food and insects. Providing opportunities for foraging helps satisfy their natural instincts.

- **Dust Bathing:** Dust bathing is essential for chickens to keep their feathers clean and free from parasites. Chickens will dig a shallow hole and roll in the dust, shaking it onto their feathers.

- **Roosting:** Chickens have a natural instinct to roost at higher points, such as tree branches or elevated roosts. Providing roosting bars in the coop allows them to perch comfortably at night.

- **Social Hierarchy:** Chickens have a pecking order or social hierarchy within the flock. Establishing this hierarchy is a natural behavior for them.

- **Egg Laying:** Nesting and egg-laying are instinctive behaviors. Providing nesting boxes with clean and comfortable bedding encourages hens to lay eggs in designated areas.

Importance of Enrichment

Enrichment activities are crucial for preventing boredom and behavioral issues in chickens. Enrichment not only keeps them mentally stimulated but also promotes physical activity and reduces stress. Enrichment is especially important when chickens are confined to a coop or run. Without enough mental and physical stimulation, chickens can develop negative behaviors like feather pecking or aggression.

Enrichment Activities for Chickens

Here are some enrichment activities to keep your chickens engaged and happy:

1. Provide a Dust Bath Area:

- Create a designated area filled with dry dirt, sand, and diatomaceous earth for dust bathing. This helps keep their feathers clean and serves as a natural form of pest control.

2. Scatter Treats and Forage:

- Scatter treats or kitchen scraps in the coop or run to encourage natural foraging behavior. This keeps them active and mentally engaged.

3. Hang a Mirror:

- Hang a small, unbreakable mirror in the coop. Chickens are curious and may enjoy pecking at their reflection.

4. Install Swings or Perches:

- Install swings or additional perches in the coop to provide opportunities for physical exercise and mental stimulation.

5. Introduce Food Puzzle Toys:

- Use food puzzle toys to challenge your chickens and make mealtime more engaging. These toys can be made from PVC pipes, containers with holes, or hanging cabbage heads.

6. Play with a Ball:

- Place a large, durable ball (such as a soccer ball) in the coop or run for chickens to peck and play with.

7. Offer Greens and Fodder:

- Provide fresh greens and fodder, like sprouted grains, for chickens to peck at during the day.

Safety Considerations

When providing enrichment activities, safety is paramount. Here are some safety considerations:

- Ensure all enrichment items are non-toxic and chicken-safe.

- Regularly inspect toys and accessories for signs of wear or damage.

- Avoid small objects that could be swallowed or cause choking hazards.

Common Behavioral Issues and Solutions

Understanding and addressing common behavioral issues in chickens can help maintain a harmonious and content flock. Here are some behavioral issues and potential solutions:

Behavioral Issue	Possible Causes	Solutions
Feather Pecking	Boredom, overcrowding, stress	Provide enrichment activities, more space, and reduce stressors.
Aggression	Establishing pecking order	Ensure sufficient space and provide multiple feeding and watering stations.
Egg Eating	Nutritional deficiencies	Provide a well-balanced diet and collect eggs frequently.
Nest Box Preference	Inadequate nesting boxes	Ensure clean, private, and comfortable nesting boxes.
Cannibalism	Boredom, overcrowding, stress	Improve coop and run conditions, and provide enrichment activities.

Keeping Chickens Entertained During Winter

Winter months can limit outdoor activities for chickens, but there are still ways to keep them entertained and happy. Here are some ideas for winter enrichment:

Winter Enrichment Ideas	Description
Hanging Treat Dispensers	Hang treat dispensers filled with corn or seeds.
Sprouted Grains	Offer sprouted grains as a nutritious winter treat.
Cabbage Piñata	Hang a cabbage head from the ceiling as a pecking toy.
Warm Grit or Oatmeal Mash	Serve warm grit or oatmeal mash for added comfort.

Observing and Understanding Chicken Body Language

Understanding chicken body language helps you gauge their well-being and emotional state. Here are some common chicken body language cues:

- **Relaxed and Content:** Relaxed posture, comfortable preening, and gentle vocalizations indicate contentment.

- **Alert and Curious:** Erect posture, attentive eyes, and quick movements show curiosity and alertness.

- **Stressed or Unwell:** Fluffed feathers, hunched posture, decreased activity, and wheezing might indicate stress or illness.

- **Aggression:** Raised neck feathers, flogging, and aggressive pecking are signs of dominance or territorial behavior.

- **Broodiness:** Puffed-up feathers, staying in the nesting box, and growling when approached signal broodiness.

The Benefits of a Happy and Enriched Flock

Keeping your chickens happy and enriched benefits both the birds and the chicken keeper. Here are some advantages of a content flock:

- **Healthy and Productive:** Happy chickens are more likely to be healthy and produce higher-quality eggs.

- **Reduced Behavioral Issues:** Enrichment activities help reduce negative behaviors like feather pecking and aggression.

- **Bonding with Chickens:** Spending time providing enrichment allows you to bond with your chickens.

- **Educational Experience:** Observing chicken behavior and responding to their needs is an enriching educational experience.

<u>Keeping Records and Observations</u>

Maintaining records and observations of your flock's behavior and responses to enrichment activities can be valuable. Here's what to record:

- **Daily Observations:** Note any changes in behavior, appetite, or egg production.

- **Enrichment Success:** Keep track of which enrichment activities your chickens enjoy the most.

- **Health Records:** Record any illnesses, treatments, or vaccinations.

In conclusion, understanding chicken behavior, providing enrichment, and addressing behavioral issues are essential for a happy and healthy flock. By offering engaging activities, observing body language, and keeping records, you can create a fulfilling environment that fosters the well-being of your chickens.

BREEDING AND INCUBATION

Breeding chickens can be a rewarding experience, whether for expanding your flock or hatching chicks for specific traits or breeds. Understanding the basics of chicken breeding, selecting suitable breeding stock, and managing the incubation process are essential for successful and responsible breeding. In this chapter, we will explore the key considerations for chicken breeding and the steps involved in incubating eggs.

Selecting Breeding Stock

Breeding starts with selecting the right breeding stock, which should be healthy, genetically diverse, and exhibit desirable traits. Here are some factors to consider when selecting breeding chickens:

- **Health and Vitality:** Choose chickens free from genetic defects, diseases, and deformities.

- **Temperament:** Select birds with friendly and calm temperaments, as these traits can be passed on to offspring.

- **Conformation:** Look for chickens with good body conformation and well-developed traits specific to their breed.

- **Egg Production:** If breeding for egg-laying ability, choose hens that consistently lay large, well-shaped eggs.

- **Meat Quality:** For breeding meat birds, select chickens with ample muscle development and good meat-to-bone ratio.

<u>Breeding Methods</u>

There are two primary breeding methods: natural mating and artificial insemination.

- **Natural Mating:** Allowing chickens to mate naturally in the flock. One rooster can typically handle 8-10 hens.

- **Artificial Insemination:** A more controlled method where semen is collected from the rooster and manually deposited into the hens' reproductive tract.

<u>Incubation Process</u>

Incubating eggs involves creating a controlled environment that mimics a mother hen's natural brooding conditions. Here are the steps for successful egg incubation:

1. Collection and Storage:

- Collect fertile eggs from selected breeding stock.

- Store eggs in a cool and humidity-controlled environment with the pointed end down.

2. Setting Up the Incubator:

- Clean and disinfect the incubator before use.

- Place a calibrated thermometer and hygrometer inside the incubator to monitor temperature and humidity.

3. Incubation Conditions:

- Set the incubator temperature to 99.5°F (37.5°C) for chicken eggs.

- Maintain humidity at 50-55% during incubation and increase it to 65-75% during the final days before hatching.

4. Egg Turning:

- Eggs should be turned at least three times a day until day 18 of incubation. This can be done manually or using an automatic egg turner.

5. Candling:

- On day 7 and day 14 of incubation, use a candler to examine the eggs for signs of fertility and embryo development.

6. Lockdown Period:

- On day 18, stop turning the eggs and increase humidity for the hatching process.

7. Hatching:

- Chicks will start to hatch on day 21. Avoid opening the incubator during hatching, as it can disrupt the process.

8. Brooding the Chicks:

- Once the chicks have hatched, transfer them to a brooder with a heat source, food, and water.

Managing Broody Hens

If you have broody hens that show a strong desire to hatch eggs, you can let them do the work of incubation and brooding. Here's how to manage broody hens:

- Provide a separate nesting area with privacy for the broody hen.

- Mark the eggs you want her to hatch to prevent adding additional eggs later.

- Allow the broody hen to incubate the eggs for 21 days or until hatching occurs.

- Once the chicks hatch, leave them with the broody hen for at least a week before introducing them to the rest of the flock.

Genetic Considerations

Understanding basic genetics can help you predict the traits and characteristics of offspring. Learn about dominant and recessive traits, and research the specific genetics of the breeds you are breeding.

In conclusion, breeding and incubating eggs require careful selection of breeding stock, proper incubation procedures, and attention to genetic considerations. By following these guidelines, you can successfully hatch and raise healthy chicks and continue to improve and expand your flock.

CHICKEN COOP EXPANSION AND FREE-RANGE MANAGEMENT

As your flock grows and you gain experience in chicken keeping, you might consider expanding your chicken coop or allowing your chickens to free-range. In this chapter, we will explore the considerations for expanding the coop, managing a free-range system, and ensuring the safety and well-being of your chickens in an expanded environment.

Coop Expansion Planning

Expanding your chicken coop requires careful planning to accommodate the growing flock comfortably. Here are some key steps to consider:

Coop Expansion Considerations	Description
Assessing Space Requirements	Evaluate the space needed per chicken based on the breed and size.
Materials and Construction	Choose durable and predator-proof materials for the expansion.
Adequate Ventilation	Ensure proper ventilation to maintain good air quality inside the coop.
Nesting Boxes and Roosts	Provide enough nesting boxes and roosts for all the chickens.
Access to the Run Area	Create easy access from the coop to the outdoor run area.

Free-Range Management

Allowing chickens to free-range can provide them with additional space to roam, forage, and exhibit natural behaviors. Here are some tips for successful free-range management:

Free-Range Management Tips	Description
Fencing and Predators	Ensure the free-range area is securely fenced to protect against predators.
Supervised Free-Ranging	Supervise chickens during free-range time to prevent escapes or disturbances.
Forage and Supplements	Provide a balanced diet, but allow chickens to forage for insects and plants.
Return to Coop at Night	Train chickens to return to the coop at dusk for safety from predators.

Managing Multiple Flocks

As your flock expands, you might have multiple age groups or breeds. Here's how to manage multiple flocks:

- **Separate Coops:** If you have different age groups or breeds, consider providing separate coops to prevent aggression and promote individualized care.

- **Introducing New Birds:** When introducing new chickens to an existing flock, use a gradual integration method to reduce stress and avoid conflicts.

- **Quarantine Procedure:** Always quarantine new birds before integrating them with your main flock to prevent disease transmission.

Seasonal Considerations

Seasonal changes can impact your chickens' well-being. Here are some seasonal considerations:

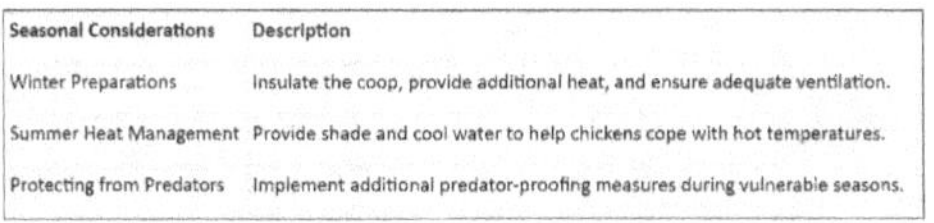

Seasonal Considerations	Description
Winter Preparations	Insulate the coop, provide additional heat, and ensure adequate ventilation.
Summer Heat Management	Provide shade and cool water to help chickens cope with hot temperatures.
Protecting from Predators	Implement additional predator-proofing measures during vulnerable seasons.

Record-Keeping and Observations

Keeping records and making observations are essential for effective coop expansion and free-range management. Here's what to record:

- **Coop Expansion Progress:** Document the process of coop expansion, including materials used and any modifications made.

- **Free-Range Behavior:** Observe the behavior of chickens during free-range time to assess their well-being and interactions.

- **Health and Egg Production:** Keep track of individual chicken health and egg production to detect any potential issues.

In conclusion, expanding your chicken coop and managing a free-range system require thoughtful planning and consideration. By providing a safe and enriching environment for your chickens, you can create a thriving and content flock that will continue to bring joy to your backyard.

Dealing with Common Challenges

Chicken keeping comes with its fair share of challenges. In this chapter, we will address common issues that chicken owners may encounter and provide practical solutions to overcome them. From dealing with pests and diseases to managing behavioral problems, being prepared for these challenges will help you maintain a healthy and happy flock.

Pest Management

Pests can be a nuisance and pose health risks to your chickens. Here's how to effectively manage common pests:

Pest Management Strategies	Description
Parasite Control	Regularly inspect and treat for external and internal parasites.
Rodent Prevention	Implement measures to prevent rodents from accessing chicken feed and eggs.
Predator Deterrence	Secure the coop and run with predator-proof fencing and lock

Disease Prevention and Control

Preventing and controlling diseases is essential for maintaining a healthy flock. Here's how to manage common chicken diseases:

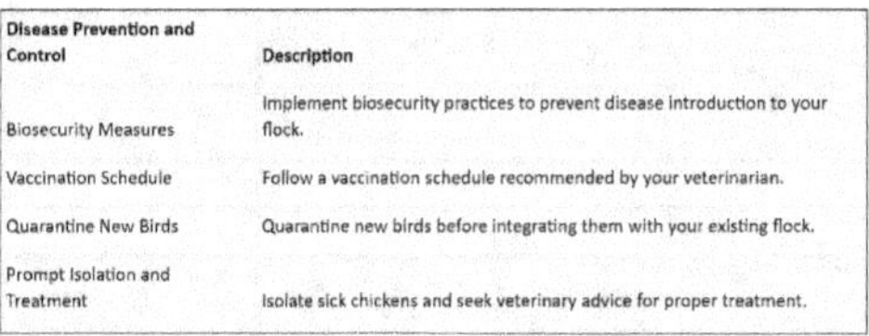

Disease Prevention and Control	Description
Biosecurity Measures	Implement biosecurity practices to prevent disease introduction to your flock.
Vaccination Schedule	Follow a vaccination schedule recommended by your veterinarian.
Quarantine New Birds	Quarantine new birds before integrating them with your existing flock.
Prompt Isolation and Treatment	Isolate sick chickens and seek veterinary advice for proper treatment.

Addressing Behavioral Problems

Behavioral problems can arise in chicken flocks. Here are solutions to address common issues:

Addressing Behavioral Problems	Description
Feather Pecking	Provide enrichment activities and a stress-free environment to reduce pecking.
Egg Eating	Collect eggs frequently and provide a balanced diet to discourage egg-eating.
Aggression and Bullying	Separate aggressive birds and reintegrate them once they have calmed down.
Broodiness Management	If broodiness becomes problematic, break broody hens using a broody pen.

Egg Production Challenges

Maintaining consistent egg production can sometimes be challenging. Here's how to address common egg production issues:

Egg Production Challenges	Description
Seasonal Egg-Laying	Expect a decrease in egg production during winter months due to reduced daylight.
Molting Period	During molting, chickens may stop laying eggs temporarily. Provide adequate nutrition.
Age-Related Decline	Older hens may lay fewer eggs. Consider replacing them with younger birds.
Stress and Health Issues	Monitor for signs of stress or health problems that may affect egg-laying.

Addressing Broody Hens

Broodiness in hens can disrupt egg production. Here's how to manage broody hens:

- Provide a separate broody pen with minimal bedding and no nesting material to discourage nesting.

- Isolate broody hens from the nesting area and remove any eggs they attempt to incubate.

- Break broodiness by gently moving the hen and providing a more stimulating environment outside the pen.

In conclusion, being prepared to address common challenges in chicken keeping is essential for maintaining a thriving flock. Implementing pest management strategies, disease prevention measures, and effective solutions to behavioral issues will help you create a healthy and happy environment for your chickens.

SUSTAINABLE PRACTICES AND CONCLUSION

In the final chapter, we will explore sustainable practices for chicken keeping and reflect on the journey of raising chickens in your home. Sustainable practices not only benefit the environment but also contribute to the overall well-being of your flock. Let's delve into eco-friendly strategies and conclude this guide with a reflection on the joys and rewards of keeping chickens.

Sustainable Chicken Keeping

Sustainability is an important aspect of responsible chicken keeping. Here are some sustainable practices you can implement:

Sustainable Chicken Keeping	Description
Backyard Composting	Compost chicken manure and bedding to create nutrient-rich soil for your garden.
Garden Integration	Allow chickens to free-range in the garden to help with pest control and fertilization.
Rainwater Collection	Collect rainwater for use in chicken coop cleaning and watering garden plants.
Natural Pest Control	Encourage natural pest control by attracting beneficial insects to the garden.

Reflecting on Your Chicken Journey

Keeping chickens is a fulfilling experience that offers numerous rewards. Take a moment to reflect on your chicken journey:

- **Lessons Learned:** Consider the lessons you've learned about chicken behavior, care, and health.

- **Connection with Nature:** Reflect on the joy of connecting with nature through caring for your chickens.

- **Eggs and Food Production:** Appreciate the fresh and nutritious eggs your chickens have provided.

- **Companionship:** Recognize the companionship and entertainment your chickens have brought to your life.

<u>Sharing the Joy of Chicken Keeping</u>

If you've enjoyed your chicken-keeping experience, consider sharing your knowledge and passion with others:

- **Educate Others:** Share your chicken-keeping journey with friends, family, and the community.

- **Assist New Chicken Keepers:** Offer guidance and support to new chicken owners to help them succeed.

- **Promote Sustainable Practices:** Advocate for sustainable and ethical chicken-keeping practices in your community.

<u>Continual Learning and Improvement</u>

Chicken keeping is a journey of continual learning and improvement. Keep seeking knowledge and applying best practices to provide the best care for your flock.

CONCLUSION

As you conclude this guide, remember that chicken keeping is a rewarding and enriching experience. By providing a nurturing and sustainable environment for your flock, you are not only benefiting the chickens but also contributing positively to the world around you.

Always cherish the moments spent with your feathered companions and take pride in the contribution you make to their well-being. The joy of collecting fresh eggs, observing their behaviors, and sharing the company of these delightful creatures will continue to bring happiness and satisfaction to your life.

Thank you for embarking on this chicken-keeping journey with us. May your chicken-keeping adventure be filled with delight, learning, and the warmth of a happy and healthy flock.